GRASS
MIRACLE OF THE EARTH

David Campbell Callender

AKA Ruth Finnegan

CALLENDER NATURE SERIES 6

Third Edition 2023

Callender
Press

Hearing
Others'
Voices

David Campbell Callender was an Irish naturalist living in Derry who was left to manage his failing family business, yet traced the flight of birds, saw the rising moon and the stars, looked to the hills and watched the grass grow beautiful under his feet.

This book is written in his honour and memory by his granddaughter Ruth Finnegan.

GRASS, MIRACLE FROM THE EARTH

ISBN 978-1-739893-71-2

First published by Balestier Press in 2019

Second Edition published by Callender Press 2021

Third Edition 2023

Callender Press I callenderpress.co.uk I Milton Keynes
United Kingdom

Covers, design and production by John at mapperou.com

Contents

Foreword

When I think of grass, I think of parks, pasture and football pitches. I don't think of rice fields, bamboo or sugarcane. I don't think of clothing, furniture or beer. This book has opened my eyes to a new way of looking at grass.

An extraordinary family of more than 10,000 grass species, it can be traced back some 50 million years. It is a family found in rain forests, deserts and mountains, from the Antarctic to the Sahara. It is a family on which most of the world's population depends for a major portion of its diet: some 70% of agricultural crops are grasses.

In an easy-to-read, well-informed style, David Campbell Callender introduces the reader to the world of grass, and its uses from agriculture to medicine, cooking to construction; its association with mythology, symbolism and the arts; its importance to our wellbeing and well just so much more.

After reading this book, I am quite sure that next time you eat a bucket of popcorn, or chop up a piece of lemon grass for a Thai curry, you will stop just for a moment and contemplate on one of the most extraordinary plant families on the planet – and the significance of the colour green. I know I will.

Hilary Macmillan

Consultant Head of Communications

Vincent Wildlife Trust

The Blue Marble: *an image of Earth taken on December 7, 1972, by the Apollo 17 crew Harrison Schmitt and Ron Evans from a distance of about 29,000 kilometers (18,000 miles) from the planet's surface.*

If we had a keen vision of all that is ordinary in human life, it would be like hearing the grass grow or the squirrel's heartbeat, and we should die of that roar which is the other side of silence. **George Eliot**

To us also, through every star, through every blade of grass, is not God made visible if we will open our minds and our eyes. **Thomas Carlyle**

**Earth from space:
was it from there that the seeds of grass first came?**

Preface

I have no idea what started me on this book. But I'm glad I did and hope you will be too.

It's true that I'd been walking on grass and seeing it and touching it just about every day of my life. But I had just never thought about it – about how it is everywhere, grows back whatever you do, delights us – even in a way holds our beautiful planet together, a foundation for nature and art and humanity.

Once started I couldn't stop. Grass, that humble weed, is amazing.

Even it's evolution is fascinating – not 'simple' at all as I'd thought – and covers hundreds of different species, including (could you have guessed?) bamboo and sugarcane, and we couldn't live or feed ourselves without it. And then there is all the art and symbolism and poetry around grass in the imagination of our thoughts.

Where could it have come from, this miraculous part of earthly life? We may never learn the answer to that mystery, but at least we can track some of its adventures.

This volume aims to give some kind of introduction to the many many dimensions of this miraculous weed of ours. For this reason, none of the accounts can go very deep and many aspects remain to be uncovered (it left to you, if you so wish, to winkle out further information from the many sources listed at the end and elsewhere).

Despite its introductory nature however I hope that you enjoy reading this story as much as we have both enjoyed discovering it.

DCC

Enjoying walking past the grass in one of 'The lungs of London'

Leading to the Taj Mahal

Grass? You must me joking!

No, no joke

Don't you see it and tread on it most days of your life? Find it all the over the place?

It clothes our planet. And brilliant in its beauty, it feeds our art and our senses and our imagination. You will see.

Grass

So what *is* grass? How did it get here and when? By some miracle, perhaps, from outer space? By (the equal miracle of) evolution? Just, somehow, naturally, from the ground? Sustained by human endeavour or earliest animals' tread?

Do we know?

Yes. Quite a bit. Read on.

Grass is the most prolific and arguably the most useful of all earth's plants, here from (nearly – you will see) as far back as we can see, even perhaps (or perhaps not…) the sustenance of dinosaurs. Certainly with incredibly ancient roots.

So – surely one of the most primitive of living things?

Not so. Grass – the great family of grass – is, you may be surprised to learn, one of the most highly evolved and enduring of earth's creations and comes in thousands of different kinds and places.

It grows around modern houses, ancient monuments, Elizabethan cottages, Greek temples and more. It is the green space that sets apart the buildings for us to view and approach through their magical liminal grass-defined space. In London parks too grass is predominant, making them, as William Pitt, the Earl of Chatham (1708-1778), was purported to claim, 'The lungs of London.'

Not just London either. Worldwide.

And much more too, for grasses are among the most versatile plant life-forms on earth. There are grain grasses, lawn and sports-field grasses, agricultural grasses, leaf and stem crops. Ornamental grasses too. Just think how many blades of grass that will be at any one time...

We maybe seldom consciously think about grass or wonder about its nature or history. But there it is all the time, just growing away.

Studland Bay, Dorset

In front of a 17th-century English cottage (Photo © Jim Graham)

So, as I say, read on for a glimpse of – yes, grass and grasses, the continuing ever-growing ever-reviving, ever the same but ever changing, miracle on this earth. A thing that is cut and cropped and burned and torn away and mutilated and nibbled and eaten but still comes back time after time after time – still there. How is that not a kind of miracle?

The grass family is indeed of astonishing importance for humans. Most people on earth rely on varieties of grass, including rice, wheat and maize (corn), for a major portion of their diet. Domestic animals are raised on pastured grass and diets partly or wholly of grasses. As well as that, grasses form an important part of the urban and suburban landscape of much of the world, and in many places near-define the agricultural and rural.

In fact, human cat owners, understanding that their pets enjoy chewing grass to aid digestion and hairball formation, will intentionally grow 'cat grass' – a common species of grass in the genus *Dactylis,* which is a cool-season perennial C3 bunchgrass native throughout most of Europe, temperate Asia and northern Africa (more on that later).

Grasses are ecological dominants, covering large areas of the earth's land surface and as a family include approximately 10,000 species classified into 600 to 700 genera. The grasses are included with lilies, orchids, pineapples, and palms in the group known as the *monocotyledons*, which includes all flowering plants with a single seed leaf.

So a bit of terminology to start off and set us among the proper scientists.

The name *Poaceae* was given to the grass family by John Hendley Barnhart in 1895 based on the tribe Poceae described in 1814 by Robert Brown, and the type genus *Poa* described in 1753 by Carl Linnaeus. The term is derived from the Ancient Greek πόα (póa; 'fodder').

Grass, then, is defined as: low, green, non-woody plants belonging to the grass family (*Poaceae*), the sedge family (*Cyperaceae*), and the rush family (*Juncaceae*). There are many grasslike members of other plants families, but only the approximately 10,000 or more species in the family Poaceae are true grasses.

This is a large and nearly ubiquitous family of *monocotyledonous* (that is, of flowering plants whose embryo includes just one cotyledon or leaf seed – more on that later).

It includes the cereal grasses like wheat, maize, rye, millet and of course rice, the most widely grown and the most important source of dietary carbohydrate on earth; bamboos, and the grasses of both natural grassland and the familiar cultivated lawns and pastures.

View to the Round Pond: an ornamental lake in Kensington Gardens, London

Grasses have stems that are hollow except at the nodes and narrow alternate leaves borne in two ranks. The lower part of each leaf encloses the stem, forming a leaf-sheath. With around 780 genera and around 12,000 species, *Poaceae* are the fifth-largest plant family, coming after the *Asteraceae, Orchidaceae, Fabaceae* and *Rubiaceae*.

Though they are commonly called 'grasses,' seagrasses, rushes and sedges fall outside this family. The rushes and sedges are *related* to the *Poaceae*, being members of the order *Poales,* but the seagrasses are members of a quite different order, *Alismatales*

Grass flowers – which you may not be aware of but are there – are characteristically arranged in spikelets, each with one or more florets. The spikelets are further grouped into panicles or spikes. The part of the spikelet that bears the florets is called the rachilla. A spikelet consists of two (or sometimes fewer) bracts at the base, called glumes, followed by one or more florets.

Author: Aelwyn: Creative Commons Attribution-Share Alike

A floret consists of the flower surrounded by two bracts, one external – the *lemma* – and one internal – the *palea*. The flowers are usually hermaphroditic – maize being an important exception – and mainly wind-pollinated, although insects occasionally play a role. The perianth is reduced to two scales, called *lodicules*, which expand and

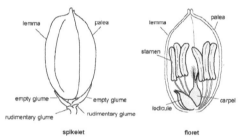

Morphology of the spikelet and floret in rice. The spikelet shows an outward form and the floret shows floral organs inside of the palea and lemma.

Source: PubMed, CC BY 3.0. Authors: Takahiro Yamaguchi, Hiro-Yuki Hirano.

contract to spread the lemma and palea; these are generally interpreted to be modified sepals – a complex structure. The fruit involves the seed coat being fused to the fruit wall.

Grass has a characteristic pattern of growth and development. Grass blades grow at the base of the blade and not from elongated stem tips. This low growth point evolved in response to grazing animals, and allows grasses to be grazed or mown regularly without severe damage to the plant.

Three general types of growth come in all grasses: bunch-type; runners (growing at the soil surface or just below ground); and rhizomes (root-like growing horizontally, spread out, at the soil surface or in various orientations underground).

C_4 Warm-season grasses like Sorghastrum nutans are highly efficient in their use of sunlight, water, and nitrogen. Photo courtesy of Hoffman Nursery.

Whatever the detail, the success of the grasses lies in part in their morphology (their basic structure) and growth processes, and in part in their physiological diversity. Most of the grasses divide into two physiological groups: C_4' grasses have a photosynthetic pathway, linked to specialized leaf anatomy that allows for increased water use efficiency, making them better adapted to hot, arid environments and those lacking in carbon dioxide

In contrast to 'C_4' 'warm-season' grasses, 'C_3' grasses are 'cool-season' grasses, thus relating the variety to their seasonal habits, as in the descriptions below:

- Annual cool-season – wheat, rye, annual bluegrass and oats

- Perennial cool-season – orchard grass, fescue, Kentucky bluegrass and perennial ryegrass

- Annual warm-season – maize, sudangrass and pearl millet

- Perennial warm-season – big bluestem, Indiangrass, Bermuda grass and witchgrass.

Grasses differ from sedges in several ways, most obviously in their sheaths and the arrangement of the leaves on the stem. In the grasses, leaf initiation begins on one side of the stem and the leaf margins grow around the stem from both sides.

C3 Turf grass: cool-season example

The root system of grass is interesting and part of the grass success story. It is fibrous and can be either rhizomatous – spread out – or (less competitive) bunch-based where the grasses remain as individual plants.

Rooting depth varies. Some species such as bluegrass have shallow rooting depths while more drought tolerant species like smooth broom have a higher proportion of their roots distributed deep within the soil profile.

Perennial grasses normally replace 25% to 50% of their roots annually. New root growth is hampered if too much leaf area is removed during the growing season but even so its survival is fabulous.

Had you noticed how, besides all this, different kinds of grasses *grow* differently? Some grow more actively and keep more leaf area near the soil surface with most of the seasonal growth as *vegetative* stems, which makes them more able to tolerate frequent grazing or clipping and still provide rapid re-growth. Other grass species, with mainly *reproductive* stems, are more sensitive to frequent clipping or grazing, thus best for hay or seasonal pasture.

That's all fine and no doubt good science, but what really *is* grass then? What do we know about it? Or rather about *them* – for as we have seen grasses come in many many different forms – and about how they work and where they came from in the first place and got started. Quite a lot as it happens, maybe more than you think. Read on...

How did grass start, and when?

So, first, what do we know about how on earth (yes, on *earth*) grass got started? A mystery? Let's begin from what we know about its history and evolution. We have to start a long way back.

We now have a fairly clear picture of the evolutionary history of the grass family based on a number or recent studies (some listed in the References at the end of this volume). Thus, the resultant evolutionary tree of the grass family has been traced from the earlier times to the present.

Grasses are flowering plants. They had to start somewhere too, but had appeared by the late Paleocene Epoch.

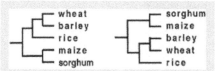

Phylogenetic trees are designed to show the relative order of speciation events. Species that are connected by a single branch point (node) are sister taxa. The more nodes separating two species, the more distantly related they are. The vertical arrangement of the names on the page does not reflect relationship and is chosen generally for convenience only. Thus, the two trees shown above represent the same evolutionary history, simply drawn differently.

Elizabeth Kellogg, Donald Danforth Plant Science Center

Simple grasslands, which bore grass but lacked the complex structural organization of sod, appeared in the Eocene, whereas short grasslands with sod appeared in the first half of the Miocene, a period of dramatic plant expansion.

We can date the origin of the grasses specifically by the appearance of grass pollen in the fossil record with, as you will see, some quite detailed analysis needed to reach the currently accepted broad conclusions.

Grasses, it seems, became widespread toward the end of the Cretaceous period, and fossilized dinosaur dung (coprolites) have been found that have been claimed (somewhat controversially) to contain phytoliths of a variety that include grasses related to modern rice and bamboo.

Fossil findings indicate that grasses evolved around 55 million years ago (MYA). Recent findings of grass-like phytoliths in Cretaceous dinosaur coprolites have pushed this date back to 66 MYA. In 2011, revised dating of the origins of the rice tribe *Oryzeae* suggested a date as early as 107 to 129 MYA. A multi-tuberculate mammal with 'grass-eating' adaptations seems to suggest that grasses were around at 120 MYA.

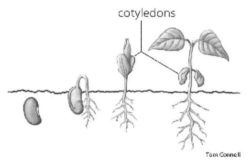

cotyledons

Tom Connell

One useful bit of evidence is that grasses and their relatives have distinctive pollen. It is nearly spherical and with a single pore. Grass pollen itself can be distinguished by minute channels or holes that penetrate the outer, but not the inner, pollen wall.

What Are Cotyledons?
Cotyledons are the food source of the plant embryo and are also the first leaves that emerge from the soil as the plant germinates. In a way, they are at first a battery or inverter and then the solar panels that energize the young plant.

The earliest firm records of grass pollen are from the Paleocene of South America and Africa, (Jacobs et al., 1999) between 60 and 55 MYA (this date is later than the major extinction events that ended the age of dinosaurs and the Cretaceous period so, unlike some earlier suggestions, it is believed now that the dinosaurs did not, after all, eat grass).

Other fossil pollen grains that may be grasses or grass relatives have been found in Maastrichtian deposits of Africa and South America (approximately 70 MYA); these were fossilized just before the end of the Cretaceous.

All this means that the ancestor of grass lived before 55 MYA but probably after 70 MYA.

By comparing grasses with their closest relatives, we can infer what sorts of changes must have happened around the end of the Cretaceous or the beginning of the Tertiary period. A major change occurred in the timing of embryo development.

This was significant as most monocotyledonous plants have largely undifferentiated embryos. Seed maturation begins after the embryo has formed a shoot apical meristem, but the differentiation of cotyledon, leaves, root meristem, and vasculature largely occurs after the seed is shed from the parent plant. In the grasses, embryo development is accelerated relative to seed maturation. (It is easy to check on technical terms unfamiliar to you through the web; some are quite interesting), but you may prefer just to get the general drift and read on).

At the same time, there was a notable change in the structure of the fruit. All the ancestors of the grasses had ovaries formed of three fused carpels, each carpel forming one locule with one ovule. In many of the close relatives, and we presume in the grass ancestors, two of those ovules abort and only one develops. In the grasses, only one locule and one ovule ever form. As the ovule develops the outer integument fuses with the inner ovary wall to form the distinctive fruit of the grasses, the grain. This structure is unique among the flowering plants.

Another striking characteristic of grasses today is how their floral and inflorescence is firm is structured. The flowers usually take the form of little spikes, or spikelets. Each of these is made up of one or more flowers plus associated bracts, usually with two stigmas and three stamens.

Outside the stamens (a relic of the three fused carpels inherited from their ancestors), the number was reduced to two after the speciation event that led to *Pharus*. The earliest species had six stamens. It is not clear precisely when the shift from six to three occurred, but it must have been after the divergence of the *Guaduella/Puelia* group.

Some of the characteristics associated with the success of the grass family evolved long before the first grass appeared in the forest. The fact that grasses as well as all their relatives are wind-pollinated, suggests that wind-pollination began millions of years before the grasses appeared on earth.

Along with wind pollination there was a reduction in perianth size and loss of pollen stickiness. All the relatives of the grasses accumulate silica somewhere in the plant so that this doubtless also originated well before the grasses themselves did. Also, a large set of monocotyledonous plants, including not only the grasses, but also the gingers, pineapples and palms, have cell walls rich in ferulic acid. Ferulic acid in the cell walls must therefore be an ancient characteristic preserved in the grasses.

And then?

Other 'grass' characteristics originated long after the first grasses appeared, notably drought tolerance and the capacity to thrive in dry open habitats. The original grasses were plants in forest margins or deep shade, characteristics that are retained today some species such as the bamboos.

These grasses apparently persisted for millions of years in such habitats without diversifying much. We still see them much as they were today. This preceded the major diversification of the family, seen in the fossil record as a marked increase in the amount of grass pollen in the mid-Miocene epoch.

All this means that we can now be fairly sure which species are most closely related.

This has produced a number of surprises, including the gradual evolution of the spikelet, the relatively late shift into open habitats, and the (apparently quite recent) diversification of the family.

Current research now places more and more grass species on phylogenetic trees (the history of their development), leading to an increasingly precise view of the order of evolutionary events.

This lays the groundwork for the main enterprise of evolutionary biology; that of understanding precisely what sorts of changes have occurred at critical junctures in evolutionary time, and therefore understanding how evolution must have worked.

Coming to the detailed history; this shows nested sets of species increasingly distantly related to the cereal crops. These certainly contain novel alleles or combinations of alleles that affect agronomically-important phenotypes.

The challenge of the future is to use the crop species as windows on the spectacular diversity produced by evolution and at the same time to use the thousands of wild grasses as tools to help understand the cereals.

So, here we are. Grass. It is a complex history. Amazing too!

So yes, we now know quite a bit about the evolutionary history of grass. It is complex indeed but – it happened. Perhaps it is still continuing? And maybe one day you too will be part of discovering yet more about this fascinating, sometimes unexpected and constantly developing subject.

Where and what is it?

So, by now, what have we? What *is* grass? Or, rather, what are *they*, for 'grass,' that simple-sounding thing, comes in many subfamilies, often strikingly distinctive.

These many different kinds generally have certain features in common. Most have round stems that are hollow between the joints, bladelike leaves, and extensively branching fibrous root systems, relatively impervious to heat and drought and to destruction by animals and humans.

Beyond that, do you have any idea how many kinds of grasses there are? All right, most people don't. Grass, low, green everywhere, is just – there…

But just look at how many kinds it comes in (yes I was surprised at first too). We have looked at some already, but now, armed with this, we can go a bit deeper and further.

Think first just about our familiar urban lawns. Here is a typical seller's advert.

Your humble lawn can be a matter of pride and maybe your sense of competition! An endless list of catchphases and ways to get your attention.

'What Type of Lawn Do I Have?'

'Choose the Right Grass for Your Lawn.'

In the United States the choice of lawn grass greatly depends on where you live. Examples are: Bahia – a tough turfgrass especially suited to the heat and humidity of the South. Bermuda: has an aggressive growth habit and gives it excellent weed resistance. Ther care many others, such as:

- Centipede
- Fescue (fine or tall)
- Perennial Ryegrass
- St. Augustine
- Buffalo
- Zoysia
- Kikuyu
- Creeping Bentgrass

Bluegrass: Kentucky or Rough-Stalked Bluegrass are the turfgrasses of choice in cooler northern areas.

The illustrations in retailers' catalogues show that contrary to most people's assumptions, there are clear contrasts between the different species. Success with your lawn, we are told, depends on many things. Not the least among them is growing the proper type of grass for your area. In general, northern areas grow cool season grasses and southern areas grow warm season varieties. Catalogues may tell us, try *Côodon dacticulata*s a perennial herb, warm season grass, or, more prosaic, Premium Green, Easy Green, Premium Shade, or Drought Buster...

Grass – big business!

Overleaf is some more information that you may (or, well, may *not*!) like to know – or maybe you or your gardening friends know it already! Additional factors such as altitude, the amount of sun or shade, the amount of foot traffic and the availability of water may affect the success of a turfgrass variety.

Grass	Mowing Height	Traffic Tolerance	Soil Type	Sun
Bentgrass	1/2-1 inches	light	tolerates acidic	full
Bluegrass	2-2 1/2 inches	light	pH 6.5-7 neutral	full
Perennial Ryegrass	2-3 inches	high	most types	full
Fine Fescue	2-3 inches	light	most types	full/ shade
Tall Fescue	2-3 inches	high	most types	full/ partial
Bahia	2-2 1/2 inches	moderate	many types	full/ mode-rate
Bermuda	1 1/2-2 inches	high	light textured	full
Centipede	1 1/2-2 inches	light	tolerates acidic	full/ partial
St. Augustine	2-3 inches	high	prefers sandy	full/ partial
Zoysia	1-2 inches	high	pH 5.5-6.5 slightly acidic	full/ partial

Here are four of the most common types of lawn grass in the UK:

Ryegrass: is the most common grass type in both the UK and the US. This is primarily due to the fact that it's both fast-growing and highly resistant, with particularly good recovery powers. Naturally, this transatlantic popularity has absolutely nothing to do with the fact that it's also cheaper than anything else by a considerable margin. That being said, ryegrass is undoubtedly an effective lawn surface; however, due to its rapid growth qualities, it does require regular mowing. This can necessitate two cuts per week during peak periods, while it also needs a lot of fertilising throughout the year to keep it on top form.

Annual Meadowgrass: is often viewed as a hybrid weed grass for its ability to infiltrate soil without an invite, meadowgrass is another common grass type found on both sides of the Atlantic. However, unlike its cost-effective cousin, this top turf has found popularity not because of its price but due to its sheer adaptability and ability to thrive in a whole host of different terrains.found in cultivated turf.

Slender Creeping Red Fescue: The slimline sibling of traditional red fescue, the aptly named slender creeping red fescue is another favoured choice amongst groundskeepers of the bowling green/golf course community. It's also found in most lawn mixes for its two-fold qualities of aesthetic appeal and durable practicality. Creeping red fescue survives well in dry and shady conditions, requiring less maintenance than most grass types in the UK. It also takes hold relatively quickly, making it a great grass choice for those starting a new lawn from scratch.

Common Brown-Top Bent: common bent (also known as brown-top bent) is – as the name suggests – extremely common in the UK, especially in moorlands and wasteland areas. While it's capable of growing on most soil types from sand to clay, common bent is most common on soils with poor acidity, requiring relatively low-maintenance. That being said, it's also capable of withstanding close mowing, making it another top candidate for bowling and golf greens.

In agriculture, the grasses used tend to be very different. For a farmer, his priorities are based on high yields, so extra grass to make hay, or for animals to graze on. Farmers tend to use different grasses for optimum yields because other species, such as Westerwolds, Italian and Hybrid ryegrass, have very big leaves. These are species you don't want in your lawn.

And beyond our familiar lawns and fields there are many other subfamilies in the great *Poaceae* grass family:

The bamboos, for instance, are evergreen perennial flowering plants in the subfamily *Bambusoideae* (the word 'bamboo' comes from the Kannada term *bambu*, entering English through Indonesian and Malay).

The *Aroisideae* subfamily separated and went off on its own fairly early in their revolutionary history and grows, unlike many other grasses, on the shaded floors of forests, tropical and warm-temperate.

Anomochlooideae: From Wikipedia

Anomochlooideae is another subfamily of the grass family. It is sister to all the other grasses and again diverged early from the rest. It includes perennial herbs that grow on the shaded floor of forests in the Neotropics.

And then sugarcanes (yes, a species of grass, and the ninth-most valuable crop worldwide) are several species of tall perennial grasses of the genus *Saccharum*, tribe *Andropogoneae*, native to the warm temperate to tropical regions of South and Southeast Asia, Polynesia and Melanesia

Saccharum officinarum, Mozambique
Ton Rulkens: Creative Commons Attribution-Share Alike 2.0 Generic

All right, perhaps you don't want to know all this. But have I now convinced you that grass is a complex subject with many ramifications? Or, at the least, that there are many many kinds of grasses – and *that* is only from having looked at the lawn varieties (there are many others – but let me not detain you with these – you can research them, if you wish, yourself).

As for *where*... that too is fascinating. As you would expect their distribution is largely controlled by environmental conditions for, despite all their hardihood, some grasses are suited to certain conditions, some to others. Temperature, soil, rainfall, human care, weeds, wind – all affect which kind flourishes best.

So grasses now grow, adapted to local conditions, in lush rain forests, dry deserts, cold mountains and even intertidal habitats. They are currently the most widespread plant type on earth and throughout the globe a valuable source of food and energy for humans and for all sorts of wildlife and organics. The grass family is in fact one of the most widely distributed and abundant groups of plants on earth. Grasses are found on every continent, including Antarctica.

Antarctic hair grass on the Antarctic Peninsula

Grasses are the dominant vegetation in many many habitats, including hillsides, plains, salt-marshes, reed swamps and steppes. They also occur as a smaller part of the vegetation in almost every terrestrial habitat dependent to an extent in local conditions.

Map of USA showing grass-type distribution by seasonal features (cool season grasses flourish in the blue zone, warm season in the light green, whereas the transition zone green – mostly has mixtures or blends of warm- and cool-season grasses)

If only large, contiguous areas of grasslands are counted as grasslands, then we can say that these cover 31% of the planet's land. Grasslands include pampas, steppes and prairies, and their grass provides for grazing mammals such as livestock, deer and elephants, as well as for many species of butterflies. Many animals ('graminivores') eat grass as their main source of food, including cattle, sheep, horses, rabbits and invertebrates such as grasshoppers and some caterpillars. Grasses are also eaten by omnivorous or even occasionally by carnivorous animals.

Because the grass meristem (the cells responsible for growth) are located near the bottom of the plant, they quickly recover from cropping at the top, hence that miracle of regeneration after cropping, cutting or mowing.

Overall we can see that grass is by now fully with us in all its diversity and age-long history and resilience. We learn it from the scientists and the historians. But most of all from our own observation, our daily lives. Again, is all that not something of a miracle, grown for us through evolution, from the earth?

What use is grass?

A lot. We all know about hay. And cows. And grazing. But what is not so widely realised is the great significance of grasses in human life. They have been cultivated as feed for people and domesticated animals for thousands of years; and for many other uses too.

First, *agriculture* is almost entirely dependent on the stalks and grains from the grass family. It provides the bulk of all feedstocks (rice, maize, wheat, barley, rye, oats, pearl millet, sugarcane and sorghum). It is true that other families like the *Fabaceae* (legumes), the *Solanaceae* or nightshade family (potatoes, tomatoes and peppers) and a few others do have some importance for us. But without grasses...

Grasses are in fact economically the most important of all flowering plants because of their nutritious grains and soil-forming function. They also have the most-widespread distribution and the largest number of individuals. Grasses provide forage for grazing animals,  shelter for wildlife, construction materials, furniture, utensils and food for humans. Some species are grown as garden ornamentals, cultivated as turf for lawns and recreational areas, or used as cover plants for erosion control.

All this is much helped by grasses' tolerance of drought and arid conditions. True they need *some* moisture to flourish. But it is truly remarkable how even the most dried-up, brown and apparently dead plants are revivified after even a small shower of rain.

After a hot dry English summer

It can grow too in the most unpromising of places. Look at the cracks between paving stones, the crevasses in rocks, the salt air by the sea. Improbable we might think. And in shifting growth and, as in sand dunes, it is grass above all, natural or humanly managed, which can hold them in place.

Thus it was a species of grass that, planted in profusion, saved a river in Irish County Donegal, from being blacked by blown sand and flooding many acres of the surrounding countryside. This was *Agrostis* (bent or bentgrass), a large and very nearly cosmopolitan genus of plants in the grass family, found in nearly all the countries in the world. It has been bred as a 'genetically modified organism' (GMO) creeping bentgrass. It is a wonderfully effective means of holding down flying sands. We could even say, more grandly, that it is the roots and blades of grass that hold the surface of our planet earth and keep it stable.

Some grasses play a part in medicine too, small perhaps but in their way vital. Volatile oils from lemon grass for example are used in traditional medicine for the treatment of various diseases. The freshly-sprouted first leaves of the common wheat plant are popularly consumed as a dietary supplement – often juiced – for their nutritious content.

Even more than all that, grass soothes and makes possible our everyday living. It provides a soft tread for feet and hooves, gentle falling ground for autumn leaves, for apples and for rain, a place for recreation and relaxation, a background and growing area for fields and gardens, an arena of exertion and heroism for grass-mowers and for sportspeople.

In human terms, grasses are the most economically important plant family of all. Their economic importance stems from several areas, including food production, industry and lawns. They have been grown as food for domesticated animals for up to 6,000 years and the grains of grasses such as wheat, rice, maize and barley have been the most important human food crops.

Grasses are also used in the manufacture of thatch, paper, fuel, clothing, insulation, timber for fencing, furniture, scaffolding and construction materials, floor matting and sports turf. And prime, wanted or unwanted, in gardening.

Of all crops grown (agriculturally, crops refers to cereals, legumes, vegetables and fruit), 70% are grasses. Three cereals – rice, wheat and maize – provide more than half of all calories consumed by humans. Cereals constitute the major source of carbohydrates for humans and perhaps the major source of protein, including rice (in southern and eastern Asia), maize (in Central and South America), and wheat and barley (in Europe, northern Asia and the Americas).

Sugarcane is the major source of sugar production. Additional food uses of sugarcane include sprouted grain, shoots, and rhizomes, and in drink they include sugarcane juice and plant milk, as well as rum, beer, whisky and vodka. Lemongrass is also used as a culinary herb for its citrus-like flavor and scent.

Wheatgrass (left) and Lemongrass (right)
CC BY 2.0, https:commons.wikimedia.org/w/index.php?curid=189911

Wild creatures profusely feed on grass – birds, animals, insects, grubs and of course some of the humble caterpillars that develop into our beautiful butterflies.

Though grass in its grain form is a valuable human food and great nutritious resource for many animals, the concept of humans having to eat ordinary common grass is, on the other hand, viewed with abhorrence, and indeed human stomachs are mal-adapted for this.

So saying that someone is eating grass or trying to is to portray them as reduced to dire straits, symbolically equating the feeder to animals (remember Nebuchadnezzar in the Old Testament, who "ate grass like an ox" (Daniel 4:33). And who can forget the harrowing descriptions of emaciated green-mouthed children during the Irish Potato Famine of 1845-49.

Many species of grass are humanly planted and grown as pasture for foraging or as fodder for prescribed livestock feeds, particularly in the case of cattle, horses, and sheep. Such grasses may be cut and stored for later feeding, especially for the winter, in the form of bales of hay or straw, or in

silos as silage. Straw (and sometimes hay) is also often used as bedding for animals. And then there are the many uses of straw. This is widely utilised and has been for many centuries or more by wild creatures of many kinds for their nests and beds, and, still, by humans for human and animal bedding, for insulation, for fodder, fuel, drinking, roofing and basket making.

Grasses are also the raw material for a multitude of purposes, including construction (remember the biblical and now proverbial near-impossibility of making 'bricks without straw'), in the composition of building materials such as cob, for insulation, and in the manufacture of paper and board such as Oriented structural straw board.

Grass fibre can be used for making paper and for biofuel production. Bamboo scaffolding is able to withstand typhoon-force winds that would break steel scaffolding. Larger bamboos and *Arundo donax* have stout culms that can be used in a manner similar to timber, *Arundo* is used to make reeds for woodwind instruments, and bamboo is used for making innumerable implements and furniture. Grass (common reed) is important for thatching, and grass roots stabilize the sod of sod houses. Reeds are used in water treatment systems, in wetland conservation and land reclamation.

Grasses also provide an important means of erosion control and stabilisation in shifting sand dunes and along roadsides, especially on sloping areas. Marram grass, which is closely related to bentgrass and beachgrass, originates on the windswept sandy coastal regions of North America, where its success in supporting and protecting against coastal erosion led to its introduction to other weather-beaten parts of the world.

Similarly, the machair that inhabits the delicate ecosystem of the coastal areas of northwest Scotland and Ireland, is quite literally the first (or last) line of defense against erosion. Although the word machair derives from the Gaelic for fertile grassy plains, it is now commonly used in scientific studies to refer to the extensive dune grasslands.

Machair east of Uig Bay, Lewis. Various lochs drain into Uig Bay cutting through the machair as they go. In the distance is Suaineabhal. By Les Ellingham, CC BY-SA 2.0, https://commons.wikimedia.org/w/index.php?curid=9232599

Forest Green Rovers are a football (soccer) club based in Nailsworth, Gloucestershire, England. They are the world's greenest football club and are planning to build a new stadium called Eco Park. The new stadium will be constructed from wood and its grass pitches will be organic – food served at the stadium will be vegan.

Grasses are also, as you have seen (and probably knew already) the primary plants for lawns, which themselves derive from grazed grasslands in Europe. Grass lawns and pitches are also, as we know, an important covering of playing surfaces in many sports, including football (soccer), American football, rugby, hockey, golf, cricket, softball, baseball and bowls (played on a bowling green – more on that later). Grass surfaces are also sometimes used for horse racing and athletic games. And even when in some settings replaced by artificial surfaces such as AstroTurf, grass, tended and looked after with immense care, remains the surface of choice in several high-profile events.

And in sport, everywhere. Among the world's most prestigious court for grass tennis is Centre Court at Wimbledon, London that hosts the final of the annual Wimbledon Championships in England, one of the four international Grand Slam tournaments. In tennis, for instance, grass is grown on very hard-packed soil, and the bounce of a tennis ball may vary depending on the grass's health, how recently it has been mowed, and the

wear and tear of recent play. The surface is softer than hard courts and clay (other tennis surfaces), so the ball bounces lower, and players must reach the ball faster resulting in a different style of play, which may suit some players more than others.

Grass surfaces are popular in numerous recreational activities also. If you look around a park or village green, you may observe people exercising pets, bird watching, reading or meditating, practising yoga and Tai Chi, or simply hanging out with friends and family enjoying the fresh air and maybe also a picnic or barbeque.

Ornamental grasses, like the perennial bunch grasses, are used in many styles of garden design for their foliage, inflorescences, seed heads. They are often used in natural landscaping, xeriscaping and slope stabilization in contemporary landscaping, wildlife gardening and native plant gardening.

commons.wikimedia.org/wiki/

And then there are the economically important grasses:

Grain crops	Ornamental grasses (Horticultural)	Leaf and stem crops	Lawn grasses
Barley	Calamagrostis	Bamboo	Bahia grass
Maize	Cortaderia	Marram grass	Bentgrass
Oats	Deschampsia	Meadow grass	Bermuda grass
Rice	Festuca	Reeds	Buffalograss
Rye	Melica	Ryegrass	Centipede grass
Sorghum	Muhlenbergia	Sugarcane	Fescue
Wheat	Stipa		Meadow grass
Millet		**Model organisms**	Ryegrass
		Brachypodium distachyon	St. Augustine grass
		Maize	Zoysia
		Rice	
		Sorghum	
		Wheat	

The primary ingredient of beer (after water) is usually barley or wheat, both of which have been used for this purpose for more than 4,000 years *(beerandbrewing.com/dictionary/UqfrcsPoAI/).*

Another more contemporary example of an everyday item produced from plant-based materials, generally corn and wheat starch, is the bio-degradable bag used to dispose of items such as food and animal waste.

In some places, particularly in suburban areas, the maintenance of a grass lawn is a sign of a homeowner's responsibility to the overall appearance of their neighbourhood.

We might even link grass-maintenance to the desire for upward mobility and its manifestation in the condition of the lawn. As Virginia Jenkins, author of *The Lawn*, put it quite bluntly, "Upper middle-class Americans emulated aristocratic society with their own small, semi-rural estates."

Indeed, 'the lawn' has been one of the primary selling points of new suburban homes, subtly shifting social class designations from the equity and ubiquity of urban homes connected to the streets with the upper-middle class designation of a 'healthy' green space and the status symbol that is the front lawn.

Many US municipalities and homeowners' associations have rules requiring lawns to be maintained to certain specifications, sanctioning those who allow the grass to grow too long. In communities with drought problems, watering of lawns may be restricted to certain times of day or days of the week.

So, in some cultural settings, grass is used to help people to compete and to climb the social scale.

And, as we will see in the next chapters, and as you may already have guessed, not the least of the uses and benefits of grass are its potential for human art and imagination.

The symbolism, mythology and literature of grass

So many associations.

To start with just think of its *colour* – more significant and more intimately related to grass than we might at first imagine. The word 'green' (you knew?) comes from the Anglo-Saxon *grene* meaning (yes) 'grass' and 'grow.' It's a colour with close and distinctive ties to nature, the environment and all things to do with the great outdoors – with grass, that is, and all that that means for us. Green is everywhere. It's the most common color in the natural world, and it's second only to blue as the most common favorite color.

Much more too, for the colour green has symbolic associations all over the world. It is the colour of life, renewal, nature and energy, and it is associated with growth, harmony, freshness, safety, fertility and rebirth.

Like most things, green can have negative associations too. It is sometimes associated with money, ambition, greed, jealousy and Wall Street – dollars are 'greenbacks' and sometimes, as in China, with infidelity. We call someone who feels sick "green around the gills," and certain yellow-grey greens have a distinctly unpleasant, institutional feel to them. We link green with envy and with greed, and even the Mr. Yuck sticker, intended to warn children away from potentially hazardous chemicals, is a bright, eye-catching green.

But mostly the associations are positive, healthful and linked with sustaining the natural environment. Thus, in Western countries, green is seen as a color of happiness and freshness. It has close ties with Islam, and in China and Japan, green is the color of new birth, youth and hope.

In many religions, furthermore, green is the color associated with resurrection, regeneration and spirituality. Houses painted green may indicate that the inhabitants have made a pilgrimage to Mecca; in Iran green, blue-green and blue are sacred colors symbolising Paradise; in Japan it is the colour of eternal life, in China, green (jade) symbolises virtue and beauty, and it is the sacred color of Islam, representing respect and the prophet Muhammad. Green is the color associated with ordinary Sundays in the Catholic Church.

It is also associated with the extra-terrestrial and supernatural. 'Little green men' are aliens from another planet while the worldwide and centuries' old 'Green Man' motif (see Anderson and Hicks, 1998), more recently also known as 'Jack in the Green,' comes in sculptures, art works and supernatural associations with places such as Stonehenge and Tintagel, and other New-Age revered places, fittingly representing a figure who is a guardian of the natural environment.

Another viewpoint on him was as the voice of inspiration to the aspirant and committed artist. The sign of his presence was the ability to work or experience with tireless enthusiasm beyond one's moral capacities. According to legend,

> "He can come as a white light or the gleam on a blade of grass, but more often as an inner mood."

The colour green appears in a number of national flags, as a representation of earth, agriculture, fertility and hope, and it is used in the flags of several Muslim countries (try looking up world flags for yourself). It is also the national color of Ireland ('The Emerald Isle') where it is commonly associated with good luck, leprechauns, four-leaf clovers and Saint Patrick's Day. It is the colour of hope because of its associations with the new verdant greens of spring. Consider also, the Green Flag roadside assistance company, whose campaign slogan reads, "With us, the grass really is greener."

As all painters know there are also variations within the green range. Paler, softer mint greens especially suggest ideas of youth, inexperience and innocence, while deeper, darker greens draw out notions of success, wealth, and money. Vibrant lime green shades are tied to energy and playfulness, and deeper olive greens are seen as representing strength and endurance.

Overall though, amidst these many shades, green essentially stands for balance, nature, spring and rebirth. It's the symbol of prosperity, freshness, and development.

 The political Green Party, now increasing in strength in many countries, naturally uses green as its brand colour, building on its associations with nature, sustaining the world, and good health so that the colour is now closely allied to ecological and progressive causes.

As you have no doubt already discerned, the semiotic use of green-coloured emblems by many wildlife and conservation organisations (including Vincent Wildlife Trust, Woodland Trust and Greenpeace) subliminally signals these internalised associations.

In the same way, we associate green with vitality, fresh growth and wealth. We generally think of it as balanced, healthy and youthful. We use green in design for spaces intended to foster creativity and productivity, and we associate green with progress – think about giving a project the 'green light.'

Stonehenge, prehistoric stone circle monument, cemetery, and archaeological site located on Salisbury Plain, about 8 miles (13 km) north of Salisbury, Wiltshire, England. It was presumably a religious site and an expression of the power and wealth of the chieftains, aristocrats, and priests who had it built – many of whom were buried in the numerous barrows close by.

We are back with grass for in at least one commonly-consulted dictionary the first meaning of 'green' is "the colour between blue and yellow in the spectrum; coloured like grass or emeralds," while the immediately-following second meaning is (yes, you didn't need to guess) "covered with grass or covered with grass or other vegetation," as in "proposals that would smother green fields with development," its synonyms including "grassy" and "grass covered."

So now we have 'green spirituality,' 'green theology,' 'green writers,' 'green publishers' (and a spate of books with titles like, *It's a green thing: diary of a teenage girl* and *Catholics going green*); 'green and growing media,' and much much more, set in a 'New Age of Green.' You can check out new titles: *https://www.publishersweekly.com/pw/by-topic/new-titles/adult-announcements/article/7560-green-theology.html*.

The Bible and other origin stories begin with the creation of the earth and the dry land (and surely the miracle of grass was there to hold it together and withstand the winds and the encroaching seas). And after all, how else do we commonly picture the Garden of Eden but as clothed with the verdure – the greenness – of grass? Amidst all this, as people are asked to pray for the earth, grass is never far away.

Green is important psychologically speaking too as it is thought to help balance emotions, promote clarity, and create an overall feeling of *zen*. As the color of nature and health, green also has close ties with emotions of empathy, kindness and compassion.

In fact, the heart chakra in yogic practice is represented by the colour green. Thus green is the colour that speaks to the heart, and nostalgia and one's native land, as exemplified in the country song "Green Green Grass of Home."

What happens to your body, then, in the presence of green and therefore when you see grass? Your pituitary gland is stimulated. Your muscles are more relaxed, and your blood histamine levels increase, which leads to a decrease in allergy symptoms and dilated blood vessels, aiding in smoother muscle contractions. Recent research suggests that those who live within reach of a park or are surrounded by green spaces may be both intellectually more alive and more balanced emotionally, and it has been shown to improve reading ability and creativity. In short, green is calming, stress relieving, and – a bit paradoxically – invigorating.

The colour green has healing power. It is the most restful and relaxing colour for the human eye to view. Green can help enhance vision, stability and endurance. Green takes up more space than other colours in the spectrum visible to the human eye and is the dominant color in the natural world. It is a natural choice in interior design as an ideal background or backdrop because we as humans are so used to seeing it everywhere.

With the colour green's association with renewal, growth and hope, green can also stand for both a lack of experience and a need grow up – 'just a green youth' – but with the implication that this growth is both likely and positive, and that the greenness is a good, perhaps necessary, foundation for reaching the full potential. In keeping with this, green stands for new growth and rebirth, common in the spring season when all of the plants are coming back to life with fresh growth and life after the cold winter months.

The colour green thus affects us physically and mentally in several different ways. It is soothing, relaxing and youthful; a colour that helps alleviate anxiety, depression and nervousness. Green brings with it a sense of hope, health, adventure and renewal, as well as self-control, compassion and harmony.

For this reason, the colour green is often used to indicate safety and 'natural products' in the advertising of drugs, of cleaning services and of medical and related goods. Green is related to nature and energy, so it is also commonly used to represent and promote 'green' products or, wordlessly, the harmless nature or fresh and calming taste of something.

Because of its associations, green colours are therefore often used to promote things to do with health, the environment and claimedly all-natural products (have a look at the colours on some of these brands).

 The widely known ecology movement – it seemed only natural – adopted green as their colour, and for nutritionists 'green vegetables' are the acme of health. In fact 'green' is nowadays a frequently used and widely understood concept (and handy rhetoric too) for a range of 'natural' and widely admired commitments and actions – around processes than can be defined as recycling, environment-friendly, energy-conserving, low-impact, healthy, anti-pollution (as in, chemical and non-natural) and the like.

 Similarly, green gemstones are promoted as helping to create balance, encourage change or growth, increase feelings of hopefulness and optimism, and break the emotional demands of others. These associations are very clear in the case of the Chinese and (rather different but equally green) New Zealand jades, both, associated with healing, growth and wisdom and, often as well, like the lowly role of grass, with humility.

All this for the familiar colour that takes its name and associations both symbolically and etymologically from grass.

Grass has a true romantic appeal to us all, as in the idea of the prairie, the far grasslands, the savannah, the pampas, the grass-clothed hills of England's 'green and pleasant land' and 'this England.' So, I suspect, will it always be. And perhaps everywhere.

With all these associations grass is, unsurprisingly, a constant theme in folksong and tradition, though often, low-lying as it is, more as a subliminal background than a subject in its own right.

There are songs like the cumulative *'The grass grew all around:'*

> Oh in the woods there was a tree
> The prettiest tree
> You ever did see
> And the tree was in the ground,
> And the green grass grew all around,
> all around,
> And the green grass grew all around.
>
> And on that tree
> There was a limb
> The prettiest limb
> That you ever did see
> And the limb was on the tree
> And the tree in the ground,
> And the green grass grew all around,
> all around.
> And the green grass grew all around.

> And on that limb
> There was a branch
> The prettiest branch
> That you ever did see
> And the branch was on the limb,
> And the limb was on the tree,
> And the tree was in the ground
> And the green grass grew all around,
> all around,
> And the green grass grew all around.
>
> **And on that branch. . .**

Indeed most children, at least in the culture I come from (Ireland), and no doubt many others too, grow up with childhood memories of playing on grass of some kind, and of songs and ditties that assume a knowledge of grass and its ways and its treatment:

> One man went to mow
> Went to mow a meadow
> One man and his dog
> Went to mow a meadow.
>
> Two men went to mow
> Went to mow a meadow
> Two men, one man and his dog
> Went to mow a meadow.
>
> Three men went to mow
> Went to mow a meadow
> Three men, two men,
> one man and his dog
> Went to mow a meadow...

And so on...

Well, it's a long and wearisome job cutting the meadow grass for hay, specially in the times when this song doubtless originated, when cutting the grass for hay, necessary for livestock through the winter months, has to be carried out with the traditional implements of cutlass, sickle or scythe. I grew up knowing too that geese grazed on grass, and many of us still struggle a little – I *think* I have mastered it now – but *love* practising the tricky tongue twister (try it!). It seems to bring us close again to the green fields of our (green) youth.

> Three grey geese
> In a green field grazing
> Grey we're the geese
> And green was the grazing.
> *(Traditional)*

Grass was and is just part of living.

The associations are not always about *green* grass or the common grass we know in lawns, as *harvesting* – of hay, of straw, of grains – is celebrated in parable and song across the world in villages, fields, halls, churches and hillsides. English schoolchildren, for example, sing songs like 'Oats and Beans and Barley,' 'John Barleycorn' and 'The Hundred Haymakers.'

So, too, in other parts of the world, such as Native America, where there are great rituals and (more a little later) a complex harvest-related mythology. Grass makes its appearance in more formal literature too. In the Bible, which, believers or not, influences us all, grass is not an uncommon metaphor:

> He makes grass grow for the cattle,
> and plants for people to cultivate,
> bringing forth food from the earth.
> *(Psalm 104:14)*

> He covers the sky with clouds;
> he supplies the earth with rain
> and makes grass grow on the hills.
> *(Psalm 147:8)*

> All flesh *is* as grass,
> And all the glory of man as the flower of the grass.
> The grass withers,
> And its flower falls away,
> But the word of the Lord endures forever.
> *(Peter 1:24)*

Though grass is seen as, like humans, short-lived, there is keen awareness of its near-miraculous regeneration in the image of God sending rain "to satisfy the parched ground and make the tender grass spring up"*(Job 38:27)*, while in the New Testament, as in much other literature, corn (which, it is too easy to forget, is also a form of grass) has an important place in event and metaphor.

Common grass tends to have a somewhat humble literary role, just the accepted setting for sweeter more colourful flowers or for human action or sweeter flowers – but so often *there:*

> "I know a bank where
> the wild thyme blows,
> Where oxlips and
> the nodding violet grows,
> Quite over-canopied with
> luscious woodbine,
> With sweet musk-roses and
> with eglantine."
> *(William Shakespeare, A Midsummer Night's Dream)*

Sometimes grass is the central focus.

C-rimson beacon Sunday
L-ets the day and sky meet;
E-arly light has come out,
N-ew dawn and morn both
G-reet.

C-old wind chill passes by,
O-n twenty-ninth of January;
N-o more mist and haze,
C-ool breeze brings glory.
E-vening shadows disappear,
P-ouring rain fades away;
C-oal clouds are all gone
I-nto the dark, blue, and gray.

O-ne birthday is remembered,
    forgetting things of the past;
N-ice feeling is obvious, as you lie
    on the meadow grass.
(Bernard F. Asuncion)

And in Carl Sandburg's famous and moving poem 'Grass:'

Pile the bodies high at Austerlitz and Waterloo.
Shovel them under and let me work –
I am the grass; I cover all.

And pile them high at Gettysburg
And pile them high at Ypres and Verdun.
Shovel them under and let me work.
Two years, ten years, and passengers ask the conductor:
What place is this?
Where are we now?

I am the grass.
Let me work.
(Carl Sandburg)

The poem is not spoken by a human being, but by the grass itself and tells soldiers to "pile the bodies high" at Austerlitz and Waterloo, two famous battlefields from the Napoleonic Wars in the early 19th century. The grass says that it covers "all." Then it lists other famous battlefields – Gettysburg from the American Civil War, and Ypres and Verdun from World War 2.

Again it commands soldiers to pile the bodies high.

The grass imagines that, in the future, ordinary people will travel on trains past the battlefields and wonder what they are. Heartbreakingly, they will not remember the battles or those now covered with grass, or see signs of them on the landscape. The grass ends with the declaration: "Let me work." In other words, Let me *grow* – that is ever the 'work' of grass.

Walt Whitman called his celebrated collection of poems on nature and the human's place within it *Leaves of Grass.* The title was an intentional pun. At the time the word 'grass' was commonly used to denote works of so-called minor literature. So, in addition to the literal meaning of the title, Whitman was saying that his book was a collection of unpretentious (like the grass) literature made up of the 'grass leaves' (pages) of his book – all this, if only subliminally, adding to the emotive overtones of the concept of grass, by now perhaps deep-rooted in our subconscious.

And more directly too in his poem opening:

> A child said, What is the grass?
> Fetching it to me with full hands;
> How could I answer the child? . . .
> I do not know what it is
> any more than he.
> I guess it must be the flag of my disposition, out of hopeful
> green stuff woven.
> Or I guess it is the handkerchief of
> the Lord,
> A scented gift and remembrance designedly dropped,
> Bearing the owner's name someway
> in the corners, that we may see and remark, and say Whose?...
> *(Walt Whitman)*

And, it is grass, we know it well in both poetry and actuality, that eventually covers graves, bones, archaeological sites, the loves and hates and tragedies of the past, the mysterious buried cities and people of far-off centuries. How can it not be evocative? Indeed being covered in grass is used, time after time as an accepted metaphorical way of referring to death in both sung and written literature, the right and touching end to a story of bravery or, above all, of life and love, setting these in the perspective of like grass's long centuries of existence.

The concept, like any other, can be treated cheerfully too, of course, just a motif available in the culture to be played with by those not as yet touched by death. One example is the song (perhaps originally 'green grave'?) for a playground game in which the girl who acts as 'mother' selects one girl in turn to eliminate from the ring:

Green gravel, green gravel,
the grass is so green,
Such beautiful flowers as
ever were seen.
Oh Annie, oh Annie,
your sweetheart has fled,
He's sent you a letter
to turn round your head.
Green gravel, green gravel,
the grass is so green,
The fairest young damsel that
ever was seen.

She's neither within,
She's neither without,
She's up in the garret a-walking about.
Green gravel, green gravel,
the grass is so green,
The pretty young maidens
are plain to be seen.

Der Kinderreigen (Hans Thoma, 1884)

Oh Annie, oh Annie,
your sweetheart is dead!
They sent you a letter
to drop down your head.
Green gravel, green gravel,
the grass is so green,
The dismalest damsel that
ever was seen.

Oh Mother, oh Mother,
do you think it is true?
Oh yes, dear! Oh yes, dear!
Then what shall I do?
Green gravel, green gravel,
the grass is so green,
The pretty young maidens are
not to be seen.

We washed her, we dried her,
we rolled her in silk,
And we wrote down her names
with a gold pen and ink.
Green gravel, green gravel,
the grass is so green,
The flowers are all faded,
there's none to be seen.

Around the green gravel
the grass is so green,
The flowers are all faded,
there's none to be seen.
(Traditional)

Even that ends in a poignant note for the adult reader, and perhaps to a greater extent in the more literary treatments like Shakespeare's:

He is dead and gone, lady,
He is dead and gone;
At his head a grass-green turf,
At his heels a stone.
(William Shakespeare, Hamlet)

Grassland is the fitting setting for action in saga and fairytale as heroes gallop many miles across grassy plains or princesses dance by night in the meadows. It features as a main subject in several takes, like the Norwegian fairytale *Doll i' the Grass* and the Irish *Grassy hollow*.

It comes in novels too, often just as a taken-for-granted but evocative background but sometimes central. And with reason.

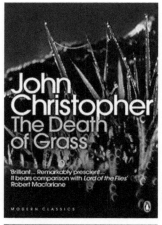

John Christopher's *The Death of Grass*, for example (a stirring title if ever there was one) shows a plague killing off all forms of grass and in so doing threatening the survival of the human species. (Can you imagine the horror of that – the end of *grass*). In fictional form the book vividly brings home to us the currently conventional and foundational place of grass in our concept of our lives and in the world.

In Ward Moore's *Greener Than You Think*, the world is slowly taken over by unstoppable Bermuda grass. Again horrifying: that something so neutral and gentle as grass could do this. . .

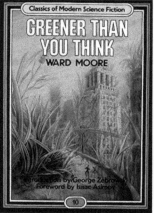

Alice Munro's article "*Save the Reaper*" by contrast draws on the idiomatic saying, "To hear the grass grow." The protagonist hears this in ways that are central to the story's significance on the topic of retelling, or rather, in an act of self-censorship, of leaving untold certain experiences of the recent past.

In fact, grass features in a number of science fiction stories and television programmes. In *Dawn of the Jedi* and *Star Wars* grass is something that grows throughout the galaxy. The planet of Alderaan, for example, was noted before its destruction for being the home of more than eight thousand species of grass; it was also the home of grass paintings. Chandrila was known for its balmgrass, a species that was soft to the touch.

Realistic depictions of believable human worlds, even in science fiction, cannot, it seems, do without the familiar presence and symbol of grass. It is everywhere.

Going on from that, many English language proverbs (and doubtless more in other languages too) build on the symbolic and metaphorical associations of grass:

> "The grass is always greener on the other side of the fence."

> "Don't let the grass grow under your feet."

> "When elephants fight, it is the grass that suffers."

You can read for yourselves here:

https://quoteproverbs.com/grass/
and https://quoteproverbs.com/green/

Although we are discussing the merits of the ever-present plant 'grass,' it should perhaps be noted that the word 'grass' has acquired some unbotanical connotations also. For example, it has entered into general parlance as a verb. To 'grass' or 'to grass up' is generally understood to refer to an informer – someone who betrays a trust to give up information. One suggested etymology is the Roman poet Virgil's metaphor, "A snake [lurks] in the grass" (1st century BC), which refers to an unethical or sneaky person, a spy perhaps – someone potentially harmful in supposedly familiar territory.

You can learn about its evolution here:

https://grammarist.com/idiom/a-snake-in-the-grass/

Another perhaps more popular theory is that 'to grass' diffused via Cockney rhyming slang – a street-smart reference to 'a grass,' as in police informer: Copper > Grasshopper > **Grass**.

While we are on the subject of slang, we should clarify that the word 'grass' denoting cannabis or marijuana is a street name only; the cannabis plant does not belong to the grass family.

Grass appears in traditional myth and legend too. One folk belief about grass is that it refuses to grow where any violent death has occurred while in Irish mythology, hungry grass (Irish: *féar gortach*; also known as fairy grass) is a patch of grass that is accursed. Anyone walking on it is doomed to perpetual and insatiable hunger.

In some stories it is the fairies who have planted the grass, possibly originating in the 1840s with the Irish Potato Famine.

An intricate example of 'fairy rings' in the Isle of Skye

And then there are fairy rings seen in the grass where it is believed fairies (in some places wolves) danced during the night; unbelievers, more prosaic, put it down to the effect of a certain type of underground fungi. Some say – can they be right? – that anyone walking through one of these rings will be struck by hunger; to safely cross through one must carry a bit of food to eat along the way (such as a sandwich or crackers, and of course some beer).

There is widespread if not always acknowledged belief in the supernatural origin of these grass rings, also known as fairy circles, elf circles, or pixie rings. They are the subject of much folklore and myth worldwide – particularly in Western Europe. While they are often seen as hazardous or dangerous places, they can sometimes be linked with good fortune.

Crop circles in cornfields are another folklore motif, with many attributing their sudden and unexplained overnight appearance to extraterrestrial sources, and many New Age groups incorporate crop circles into their belief systems.

Many stories surround them too. For instance, a 17th-century English tale called the Mowing-Devil depicts the devil with a scythe cutting a circular design in a field of oats.

The text with the image states that the farmer, disgusted at the wage his mower was demanding for his work, insisted that he would rather have "the devil himself" perform the task. The circular form showed the farmer that it had been caused by the devil himself.

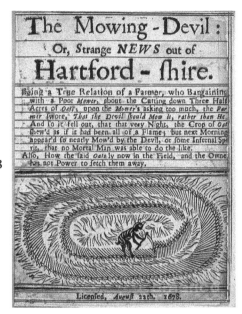

A more recent example is the 1948 German story *Die zwölf Schwäne* (The Twelve Swans). Every morning, a farmer finds a circular ring of flattened grain in his field. After several attempts to discover the cause, his son sees twelve princesses disguised as swans, who took off their disguises and danced in the field.

Today a number of people – and not just 'weirdos' either – are convinced that these crop circles are caused by ball lighting and that the patterns are so complex that they have to have been controlled by some paranormal entity such as Gaia (asking to stop global warming and human pollution), God, supernatural beings (for example Indian devas) or perhaps the collective minds of humanity through a proposed 'quantum field.'
1678 pamphlet on the *"Mowing-Devil" (Wikipedia)*

All this around our familiar growth of grass in its common forms – something, somehow, deep there?

To add to this, corn and corn spirits and deities come in the mythic, symbolic, tales of many cultures: Classical Greek, Egyptian, Hittite, Indian and more. They are central in the mythology of Native Americans, Mayan, Aztecs and others in North and Meso-America who worshiped corn gods and developed a variety of myths about the origin, planting, growing and harvesting of maize.

Most of these corn deities are female and associated with fertility. They include the Cherokee goddess Selu; Yellow Woman and the Corn Mother goddess Iyatiku of the Keresan people of the American southwest; and Chicomecoatl, the goddess of maize who was worshiped by the Aztecs of

Mexico. The Maya believed that humans had been fashioned out of corn, and they based their calendar on the planting of the cornfield.

Male corn gods do appear in some legends. The Aztecs had a male counterpart to Chicomecoatl, called Centeotl, to whom they offered their blood each year, as well as some minor corn gods known as the Centzon Totochtin, or "the 400 rabbits." The Seminole figure Fas-ta-chee, a dwarf whose hair and body were made of corn, was another male corn god. He carried a bag of corn and taught the Seminoles of Florida how to grow, grind and store corn for food. The Hurons of northeastern North America worshiped Iouskeha, who made corn, gave fire to the Hurons and brought good weather.

The Zuni people of the southwestern United States have a myth about eight corn maidens. The young women are invisible, but their beautiful dancing movements can be seen when they dance with the growing corn as it waves in the wind.

One day the young god Paiyatemu fell in love with the maidens, and they fled from him. While they were gone, a terrible famine spread across the land. Paiyatemu begged the maidens to turn back, and they returned to the Zuni and resumed their dance. As a result, the corn started to grow again.

Where did corn come from? Scientists tell us it has been grown, in a variety of forms and many places in the world for countless millennia. Other stories explain, in their own terms, how it started and became the most important food crop in Central and North America. For example, a large number of Indian myths deal with the origin of corn and how it came to be grown by humans.

Many of the tales center on a Corn Mother or other female figure who introduces corn to the people. In one myth, told by the Creeks and other tribes of the southeastern United States, the Corn Woman is an old woman living with a family that does not know who she is. Every day she feeds the family corn dishes, but the members of the family cannot figure out where she gets the food.

One day, wanting to discover where the old woman gets the corn, the sons spy on her. Depending on the version of the story, the corn is either scabs or sores that she rubs off her body, washings from her feet, nail clippings, or even her feces.

In all versions, the origin of the corn is disgusting, and once the family members know its origin, they refuse to eat it.

The Corn Woman tells the sons to clear a large piece of ground, kill her, and drag her body around the clearing seven times. However, the sons clear only seven small spaces, cut off her head, and drag it around the seven spots. Wherever her blood fell, corn grew. According to the story, this is why corn only grows in some places and not all over the world.

In another account, the Corn Woman tells them to build a corn crib and lock her inside it for four days. At the end of that time, they open the crib and find it filled with corn. The Corn Woman then shows them how to use the corn.

Other stories of the origin of corn involve goddesses who choose men to teach the uses of corn and to spread the knowledge to their people. The Seneca Nation of Indians of the northeast tells of a beautiful woman who lived on a cliff and sang to the village below. Her song told an old man to climb to the top and be her husband. At first, he refused because the climb was so steep, but the villagers persuaded him to go.

When the old man reached the top, the woman asked him to make love to her. She also taught him how to care for a young plant that would grow on the spot where they made love. The old man fainted as he embraced the woman, and when he awoke, the woman was gone. Five days later, he returned to the spot to find a corn plant. He husked the corn and gave some grains to each member of the tribe. The Seneca then shared their knowledge with other tribes, spreading corn around the world.

There are linked ceremonial events too, again bringing out the centrality of that precious grass species, corn. Native Americans of the southeast hold a Green Corn Dance to celebrate the New Year and thank the spirits for the harvest in July or August. None of the new corn can be eaten before the ceremony, which involves rituals of purification and forgiveness and a variety of dances. Finally, the new corn can be offered to a ceremonial fire and a great feast follows.

Mayan stories on the other hand give the ant – or some other small creature – credit for the discovery of corn.

In this account, the ant hid the corn away in a hole in a mountain, but eventually the other animals found out about the corn and arranged for a bolt of lightning to split open the mountain so that they could have some corn too.

By George Catlin - http://americanart.si.edu/collections/search/artwork/?id=4078, Public Domain, https://commons.wikimedia.org/w/index.php?curid=6638534

The painting by George Caltin shows the Hidatsa people of the North American Plains celebrating the corn harvest with their Green Corn Dance, the beginning of the New Year.

The fox, coyote, parrot and crow gave corn to the gods, who used it to create the first people. Although the gods' earlier attempts to create human beings out of mud or wood had failed, the corn people were perfect. However, the gods decided that their new creations were able to see too clearly, so they clouded the people's sight to prevent them from competing with their makers.

The Lakota Plains Indians say that a white she-buffalo brought their first corn. A beautiful woman appeared on the plain one day. When hunters approached her, she told them to prepare to welcome her. They built a lodge for the woman and waited for her to reappear. When she came, she gave four drops of her milk and told them to plant them, explaining that they would grow into corn. The woman then changed into a buffalo and disappeared.

There are a multitude of linked myths and ceremonies. For the Penobscot Indians, the Corn Mother was the first mother of the people. Their creation myth says that after people began to fill the earth, they became so good at hunting that they killed most of the animals. The first mother of all the people cried because she had nothing to feed her children.

When her husband asked what he could do, she told him to kill her and have her sons drag her body by its silky hair until her flesh was scraped from her bones. After burying her bones, they should return in seven months, when there would be food for the people. When the sons returned, they found corn plants with tassels like silken hair. Their mother's flesh had become the tender fruit of the corn.

Another Corn Mother goddess is Iyatiku, who appears in some Pueblo legends. She leads human beings on a journey from underground up to the earth's surface. To provide food for them, she plants bits of her heart in fields to the north, west, south and east. Later the pieces of Iyatiku's heart grow into fields of corn.

The mythology metaphorically leads, once again, into deep places of the human spirit and an insight (you doubtless have it too) into the miraculous-seeming cycling and recycling of the seasons and of human life.

One such seasonal custom performed at harvest time in the UK is the honouring of the last sheath of corn and the creation of corn dollies. These decorative forms fashioned out of cornhusks vary regionally in style and intricacy but essentially all symbolise the spirit of the corn and encourage an abundant harvest the following year. Nowadays, some people say the last sheath is kept to teach the new corn how to grow.

Yorkshire Spiral or Drop Dolly

By Renata (talk) - Public Domain, https://commons.wikimedia.org/w/index.php?curid=12738469

You can view some images and read about the long rich history here:

http://projectbritain.com/harvest/corndollies.html
https://hauntedpalaceblog.wordpress.com/2019/08/11/corn-dollies-from-the-old-crone-to-the-maiden/

In literature and song green meadows and grassy slopes are often associated with meetings of love, with memory and with natural growth. Here we leave it in the evocative background to William Butler Yeats' poem/song "Down by the sallee gardens:"

> She bade me take life easy,
> as the grass grows on the weirs;
> But I was young and foolish,
> and now am full of tears.

Added to this there can be the scent of freshly cut grass – produced, mundanely, by cis-3-Hexenal, but for all that amazingly evocative and lingering in our senses.

That and so much else is pulled back and evoked in our unconscious awareness through literature, song, myth, memory, ritual and the colour of grass.

In their quiet way dry grass flowers and seeds, cost free and abundant, are can be used for beautiful room decorations

The art and beauty of grass

Grass as such is not as prominent in visual art as one might expect from its symbolic significance and near-ubiquity throughout the world and the centuries.

This is no doubt in part because in our most familiar sightings of grass it is both low-lying and largely unnoticed.

But there is another reason too: the nature and history of the green pigment that could be used to represent grass.

Though grasses other than the common green ones and the dry forms of hay could be and were represented in paintings, the *greenness* that we tend to consider the most attractive and distinctive feature of our familiar grass was in earlier years near-impossible to capture successfully in paint.

Garden of Eden: Adi Holzer, 2012, Attribution, via Wikimedia Commons

Left panel of The Garden of Earthly Delights triptych

Hieronymus Bosch, c.1450-1516

But though in paintings grass is usually the background, sometimes grass itself is the central feature. It can be used in its dried form instead of flowers in house decorations for instance and close-up photography reveals something of its intricate natural art.

More important and widely admired, think of the beautifully-landscaped gardens round stately homes, of 'Capability Brown' creations, of Chinese palace grounds, of Versailles. In all these beautiful places grass regularly has a major role.

Grass, abundant and available, can also offer the opportunity for people to make their own art, above all in their own gardens. We see some amazing sculptures mounded and

Lost Gardens of Heligan, Cornwall

cut from grass, creative mixes of different grasses chosen to beautify a garden, figures mown out and shaped on lawns.

Here, for example, are a set (omitting all but one of the detailed how-to illustrations) of full blow-by-blow instructions for an extravagant piece of 'grass art:'

"It's easy to mow spirals, diamonds or a chequerboard into your handsome green canvas, but if you really want to show off to the neighbours, try clipping out a famous figure! You will need a second-storey balcony or scaffolding tower, so you can project the image onto the lawn, then it's time to get shearing. It may take a bit of patience, but grab a mate to help and you'll soon be appreciating the green, green grass of home.

Creating Elvis

Gather your supplies:

Black-and-white printout of your choice (we used an image of Elvis Presley).

A4 transparency sheet and A4 card in black Long-handled grass shears or Cordless electric grass shears. 3.6m-high (approx.) aluminium scaffolding tower (hired, if needed)

You'll also need: Torch; leaf blower; ear muffs; laser pointer; scissors; tape; paddle-pop sticks.

Step 1 Using a photocopier, transfer chosen image onto transparency sheet. Cut out centre of black card and discard, then tape frame to transparency.

Step 2 At night, on top of balcony or scaffolding tower, shine torch behind transparency to project image onto lawn.

Step 3 Position paddle-pop sticks along outline of image, using laser pointer to help guide placement of sticks.

Step 4 In daylight, trim between paddle-pop sticks using long-handled shears. Use electric shears for smaller, more intricate areas.

Step 5 When complete, trim edges down to ground to help highlight the figure. Wearing ear muffs, use leaf blower to remove any leaves, sticks and clippings."

And that is that (presumably some people do it).

It may end up looking like this...!

Even if in the past the greenness of grass had to lie principally in the imagination, as in the landscapes or background in mediaeval or renaissance religious paintings, various species of grass have often made a spectacular appearance in pictures, as, for example, in Vincent van Gogh's striking pictures of fields.

Harvest in Provence (1888)

Wheatfield with crows (1890)

Grass comes in contemporary film and photographs too, both of wild nature and private gardens.

Grass is also often used as an effective and evocative frame, the background or foreground to set off other subjects. This comes both in earlier art and in more recent pictures and photographs, as in the examples here.

The future...?

A great example of 'looking forward' is by The National Trust of the UK. The Trust has announced its largest ever wildflower grassland project as part of its efforts to help save threatened habitat and tackle the nature and climate crises. The Trust is transforming a 70 mile (113km) stretch of the coastline between Torridge and west Exmore into a haven for wildlife by planting pockets of wildflower grassland.

Trust rangers sowed the first 86 hectares (213 acres)(of wildflower grassland in the autumn of 2022 to attract wildlife, including voles, kestrals, bats, bees, and butterflies. The grassland will also provide nature-rich places for people to enjoy. All the areas sown recently are near public fights of way, allowing easy access.

Flower-rich grassland are very rare, with 97 per cent being lost over the century, but lowland grassland creation is an effective and realtively quick-way to improve habitas for wildlife and boost biodiversity.

Once fully established, the first grasslands will become donor seed sites for the rest of the project, which will be completed by 2030. The National Trust team has worked closely with the North Devon Area for Oustanding Natural Beauty on the project, and they've provided the Trust with funds for equipment, training and community involvement through The National Lottery Fund, UK. *(Source: NT Spring Magazine 2023)*

Yes in this humble, overlooked enduring plant so much history, so much complexity, resilience, healing, calm, symbolism, myth, emotion, beauty – holding our world and our lives together in its gentle clasp. How can all that, its history, its multiplicity, its lovely endurance holding our earth, not be a kind of miracle?

Just – the green growing grass.

Ruth

'Wild' grass, North Corfu

You might also like...

David Attenborough, *The living planet,* Collins, 1992.

Jo Chatterton, *Grasses,* Southwater, 2014.

The Nature Conservancy Council.

https://www.nature.org/en-us/about-us/where-we-work/europe/stories-in-europe/tnc-uk-foundation-limited/

Roger Phillips, *Grasses, ferns, mosses and lichens of Great Britain and Ireland,* Pan, 1980.

Kate Smith, 'Color symbolism and meaning of green.'

http://www.sensationalcolor.com/color-meaning/color-meaning-symbolism-psychology/all-about-the-color-green-4309#.W4_GWPrTWhA/

Clara-Sue Tidwell, *Native American knowledge systems,* Balestier Press, 2018.

Alan Titchmarsh, *How to garden. lawns, paths and patios,* BBC Books, 2009.

World Wildlife Fund (WWF).

https://www.worldwildlife.org/
?_ga=2.92207578.1372634282.1573655718-1887520138.1573226989

North Devon grassland land preparation | © National Trust Images/John Miller

'Coastal' grass, North Corfu

References

Michael Allaby, *A dictionary of plant sciences,* Oxford University Press, 2019 (also earlier editions)

William Anderson and Clive Hicks, *The green man,* Compass Books, 1998.

Elizabeth A. Kellogg, *Evolutionary history of the grasses,* Plant Physiology, 2001, *https://www.researchgate.net/publication/ 12088356_Evolutionary_History_of_the_Grasses*

Virginia Jenkins, *The lawn: a history of an American obsession,* Smithsonian Books, 1994.

James D. Mauseth, *Botany,* Jones and Bartlett, 6th edition, 2016.

Poaceae [Grass]

https://en.m.wikipedia.org/wiki/Poaceae

'Rural" parkland grass, Buckinghamshire

Acknowledgements

An over-arching work like the present inevitably relies on an immense range of sources, too many to list in full here. The most important are mentioned below, chief among them the wonderful expert articles in *Wikipedia* extensively relied on and both directly and indirectly quoted above - our most sincere thanks.

In addition the publishers wish to express their gratitude to my good friend John Hunt, cartographer and photographer for his work on this new edition, and to the many sources for the images reproduced here (if despite their care, they have inadvertently used any in copyright, please let me know). To the best of our knowledge, all the images that are not specifically attributed (see John Hunt's list below) come either from unsplash or from the author's collection.

© **Photos by John Hunt and Alex Hunt**

Covers.

Pages: 4,. 1§2, 14, 18, 22, 23, 27, 31, 35,. 36(2), 37, 39, 42/43 (Alex), 46/ 47, 62, 65, 66 (Alex), 70, 72, 74, 76

'Sporting' grass, Somerset

Questions for discussion, reflection and action

Who owns grass? And who should be responsible for it? (A good topic to debate).

Can you find any other poems/songs about, or literary accounts of, grass? If not, or if so, try writing another (better?) one either as yourself or as one of the animals or other creatures that feed on it. Or as grass itself.

Get your friends to do likewise, then show their efforts round and choose the best few to put on the HOV Facebook page, with a challenge to others to do likewise (you never know, if there are enough, and good ones, we might want to publish them as a collection, so be sure to include your name).

In pictures in which grass appears, do you think it adds anything to the overall impression? If so, how and why? Would it be possible to paint or photograph a landscape *without* including grass in some form or other, and if not why?

Are there controversies over grass? And if so whose side are you on?

What are your own and/or your friends' associations with the colour green? Do they agree with what is said here? Are these associations important to you (or your friends, or teachers) and if so how and why?

What would *your* worst nightmare about grass be? Why? (Worth writing a story about perhaps).

Did *you* know that geese grazed in grass? And (just for a laugh, life isn't all serious!), which one of you and your friends or family can say fastest *and right* the tongue twister about three grey geese? (Then you could always make up another one to really outwit them!).

How many of the unnamed illustrations of grass in the text can you identify? (The references at the end will help but you may need to send out a request for assistance or engage in further study. Good luck!).

You could set up an informal group to explore questions about grass that might be interesting to research (different people could, if you wished, be chosen to follow up on each, then put it all together in a book or booklet for others – lots of photographers and pictures please, and tell us about it.

Hearing Others' Voices

ABOUT HEARING OTHERS' VOICES

A transcultural and interdisciplinary series edited by anthropologist Ruth Finnegan and others for Callender Press, to inform and engage general readers, under-graduates and, above all, young adults and students to reflect on who and where they are and to explore recent advances in thought, unaccountably overlooked areas of the world, and contemporary key issues.

Each volume is by an acknowledged expert (international authority, fellow of a national academy, professor, or the like, together with the brightest of younger scholars and practitioners) – authors who are eager to communicate outside the too often closed realms of academe. General readers will find much to interest them, set out in straightforward but not simplistic terms. But it is above all to the eager young that the series is directed – the generation who will soon hold our precious earth and its resources and peoples in their hands and be responsible for it.

Less textbooks, more exciting collections for reflection and challenge, the series gives readers a unique route into greater awareness of our wonderful world, far and near, east and west, past and present.

The series logo was created specially for us by the celebrated designer Rob Janoff, creator of the Apple logo, hopefully a feature that will play well with a young adult computer-mad audience.

The first volumes were released in November 2018, preceded by the October launching Chengdu, west China of the Chinese version of Rob Janoff's amazing personal account of how he created the famous apple logo.

Hearing Others' Voices

ABOUT HEARING OTHERS' VOICES

Welcome. I look forward to receiving your ideas, questions, arguments, criticisms and challenges. Let's hear your views, read your poems and see your art and other materials. Photos and videos too, images, and links to music and poetry and thoughts please, your own and others'.

The series, after all, is called **Hearing Others' Voices** – yours very much included – so that's what it's all about.

Ruth f

ruthhfinnegan.com

RUTH FINNEGAN
OBE FBA FAFS FRAI

Emeritus Professor
The Open University

Anthropologist and
prize winning author

www.hearingothersvoices.org

callenderpress.co.uk

callendervision.org

Why not join the Oak Grove Readers and Writers Association on Facebook
and get the FREE eBook LISTEN TO THIS.

Callender Press

Ingram Content Group UK Ltd.
Milton Keynes UK
UKHW020844210423
420550UK00008B/66